Automobile

Engineering

Royal Society of Automobile Engineers

American Institute of Automobile Engineers

Automobile Engineering

This book has been prepared by the two world leaders in the field of automobiles – The Royal Society of Automobile Engineers, American Institute of Automobile Engineers.

This book would be extremely useful in preparing for certified courses in the field of Automobile Engineering. This book is the result of several years of hard work by distinguished members of the two internationally renowned professional bodies. Several experts have recommended this book as a must-have in the arsenal of the dedicated professional. This book will help you clear almost all the competitive exams with remarkable ease. Most of the leading universities around the world have recommended this authoritative book for preparing for examinations. The book contains diverse questions, starting from basics to in-depth knowledge and latest developments in this field. By preparing using the questions and assignments in this book, one can master the fundamentals, gain a sound knowledge about the basics and have a head start while beginning your professional life.

For the solution to these questions and assignments, it is strongly recommended to follow "*Automobile Engineering Vol. I and II*" by Dr. Kirpal Singh.

Automobile Engineering

Answer briefly:

1. List the types of engines.
2. What is meant by firing order?
3. What is the function of fuel injector?
4. What is the function of a carburettor?
5. What is the necessity of engine lubrication?
6. List the types of lubrication.
7. What is the function of clutch in automotive system?
8. List the type of gear boxes.
9. What is power-steering?
10. What do you mean by camber related to wheels?
11. What is lean burn engine?
12. What are the types of carburettors?

13. What are the applications of sensors in automobiles?

14. What is CRDi?

15. Name any two lubricants used in automobiles.

16. What is the function of rear axle?

17. What is a synchromesh gear box?

18. What are the functions of tyres?

19. What is meant by engine rating?

20. What are Euro-norms?

21. What are the drawbacks of conventional ignition systems?

22. What is atomizer?

23. What are the components of water cooling system?

24. What is the candle power required for the headlight of a mid-sized car?

25. What are the types of universal joints?

26. What are the functions of frame?

27. What are the limitations of a rotary engine?

28. What is meant by emission control?

29. What are the functions of metering in a carburettor?

30. What is lean mixture?

31. What are the types of electrical circuits used in a passenger car?

32. What are the requirements of a good lubricant?

33. What is overdrive?

34. What is self-locking differential?

35. What are the functions of suspension?

36. What is meant by tyre rotation?

37. What are the types of vehicles?

38. How are automobiles classified?

39. What are the various processes which influence the factors of carburetion?

40. What are the requirements of an injector nozzle?

41. List the factors that affect the life of a battery.

42. What are the advantages of an epicyclic gear box?

43. List the functions of a suspension system.

44. What are the main components of a steering system?

45. What is idling of engines?

46. Name the parts of Electronic Ignition system?

47. What is a propeller shaft?

48. What is a differential?

49. Name the different types of brakes.

Answer in detail:

1. With a neat sketch, explain the working of a turbocharged engine.

2. Write a note on the following: (i) Emission and its control (ii) Advantages of lean burn engine (iii) Euro norms

3. Discuss the following: (i) Working of a simple carburettor (ii) Electronic fuel injection

4. With a neat sketch, explain the working principle of diesel injector.

5. Explain the working of common rail injection system.

6. Draw a typical electrical circuit used in a car and explain the same.

7. Sketch and explain the pressure lubrication system

8. With a neat sketch, explain the electronic ignition system.

9. Sketch and explain splash lubrication system used in a car.

10. With a neat sketch, explain the working of a multi-plate clutch.

11. Explain the major function of automobile differential.

12. With a neat diagram, explain the working features of a hydraulic transmission system.

13. Briefly explain the principles of operation of any one type of gear box.

14. Explain the working principle of a power steering.

15. Explain the importance of suspension systems in a car.

16. Sketch and explain the front and rear end suspension system.

17. Discuss the terms toe-in and toe-out.

18. Briefly discuss the working features of a CNG engine.

19. Describe the methods used for mechanical balance and power balance.

20. What are the various types of fuel-feed systems? Briefly explain.

21. Describe idling and low speed features in a carburettor.

22. Describe the working of a Multi-Point Fuel Injection system.

23. How does the fuel injection pump in a CI engine work? Explain with a simple sketch.

24. Enumerate and briefly explain the various properties of lubricants.

25. Describe briefly the various lubricating systems with simple sketches.

26. Describe the working of a centrifugal clutch? What are its advantages and limitations?

27. Give a note on clutch troubles and its causes.

28. How do you classify automobile transmission? Describe briefly the sliding mesh gear box with neat sketches.

29. Draw and explain the general arrangement of a steering mechanism.

30. Write notes on (i) Caster and Camber (ii) Automobile wheels (iii) Types of braking systems.

31. What are the differences between internal combustion engines and external combustion engines?

32. Give the classification of internal combustion engine. Briefly discuss one type.

33. Explain firing order and its significance with illustrations.

34. Describe magneto-ignition system for a four cylinder engine.

35. Explain the working of electromagnetic clutch.

36. What are the functions of steering systems and what are their requirements?

37. Describe the working of a cam and roller steering gear with a neat sketch.

38. What are the essential parts of an automobile engine? Describe its functions.

39. What are the methods of minimizing the crank shaft torsional vibration? Discuss with examples.

40. Describe engine balancing of six cylinder engine with a neat sketch.

41. Draw the wiring circuit of the lighting system and the horn circuit of a modern car.

42. State the gear box troubles, causes and remedies.

43. What is a torque converter? Discuss with a neat sketch.

44. What are the types of steering gears? Describe worm and sector steering gear with a neat sketch.

45. Describe the working of a hydraulic brake.

46. Compare CNG with conventional automotive fuels and describe various components of the kit for converting a petrol/diesel vehicle into a biofuel vehicle with CNG as one of the fuels.

47. Discuss the various power losses that take place between the engine and the driving wheels.

48. Explain the effect of power-to-weight ratio on the performance of an automobile.

49. With the help of a curve representing the variation of mixture requirements, explain why (i) an idling engine requires a rich mixture (ii) a cruising engine requires an economy mixture and (iii) maximum power demands a rich mixture.

50. Describe the drive mechanisms of the starting motor and the over-running clutch drive.

51. Explain wet sump lubrication system and dry sump lubrication systems.

52. How can the life of a tyre be increased?

53. Explain in detail the stratified charged engine system

54. Briefly explain the different sensors used in Automobile systems.

55. Discuss the details of the fuel pumps and the fuel injection pumps.

56. Write notes on automatic transmission.

57. Write notes on the brakes used in the latest automobile systems and tyres used in automobile systems.

Space to write answers

Space to write answers

Space to write answers

Space to write answers

Space to write answers

Space to write answers

Space to write answers

www.ingramcontent.com/pod-product-compliance
Lightning Source LLC
Chambersburg PA
CBHW051829170526
45167CB00005B/2218